THE UNIQUE ASPECTS OF THE 2024 SOLAR ECLIPSE

Unveiling the Magnificence of the 2024 Total Solar Eclipse and why it could be the best eclipse for years.

Julian Anderson

INTRODUCTION TO TOTAL SOLAR ECLIPSES

Total solar eclipses are awe-inspiring celestial events that captivate humanity with their dramatic display of the cosmos in action. Occurring when the Moon passes between the Earth and the Sun, obscuring the Sun's light and casting a shadow upon the Earth, these rare phenomena have fascinated civilizations throughout history. In this section, we will delve into the definition of a total solar eclipse and explore its profound historical significance.

What is a Total Solar Eclipse?

A total solar eclipse occurs when the Moon, in its orbit around the Earth, aligns perfectly with the Sun, creating a brief period of darkness on Earth as the

Moon's shadow traverses the planet. This alignment results in the Moon completely blocking the Sun's bright disk, revealing the Sun's outer atmosphere, known as the corona, to observers on Earth. The sky darkens, stars become visible, and the temperature may drop, creating a surreal experience for those fortunate enough to witness it.

Total solar eclipses are relatively rare events, as the alignment of the Sun, Moon, and Earth must occur with remarkable precision. Despite the Moon being approximately 400 times smaller than the Sun, its proximity to Earth and the alignment of its orbit allow it to appear the same size as the Sun when viewed from our planet. This cosmic coincidence is what enables the Moon to perfectly cover the Sun during a total solar eclipse, creating one of the most captivating sights in the natural world.

Historical Significance of Solar Eclipses

Throughout history, total solar eclipses have held profound significance for various cultures and civilizations around the world. Ancient civilizations often interpreted these celestial events as omens or messages from the gods, attributing them to supernatural forces or divine intervention. The sudden darkness during the day evoked fear and awe, leading to the development of myths, legends, and rituals surrounding eclipses.

One of the earliest recorded observations of a solar eclipse dates back to ancient China, where historical texts from the Shang dynasty describe an eclipse occurring in 2136 BCE. Ancient Chinese astronomers meticulously documented solar eclipses and recognized them as predictable phenomena governed by the movements of celestial bodies. Their observations laid the foundation for early eclipse prediction methods and astronomical understanding.

In other cultures, solar eclipses were seen as portents of change or impending doom. The ancient Greeks believed that eclipses signaled the wrath of the gods, while the Maya of Central America incorporated eclipse predictions into their sophisticated calendar system. Across continents and civilizations, solar eclipses inspired awe, curiosity, and philosophical reflection, shaping our understanding of the universe and our place within it.

In modern times, total solar eclipses continue to captivate scientists, astronomers, and the general public alike. Beyond their historical and cultural significance, eclipses offer valuable opportunities for scientific research and astronomical observation. By studying the behavior of the Sun's corona, astronomers can gain insights into solar activity, magnetic fields, and the dynamics of our solar system.

In summary, total solar eclipses represent a convergence of scientific wonder, cultural mythology, and human curiosity. As we continue to explore the mysteries of the cosmos, these celestial events serve as reminders of the interconnectedness of humanity and the vastness of the universe.

THE 2024 TOTAL SOLAR ECLIPSE : OVERVIEW

The year 2024 promises an astronomical spectacle of unprecedented proportions with the highly anticipated total solar eclipse scheduled to grace the skies. In this section, we will delve into the specifics of this celestial event, including its date, time, path of totality, as well as the fervent anticipation and excitement surrounding its occurrence.

Date, Time, and Path of Totality

Mark your calendars for April 8, 2024, as this date will herald the arrival of one of nature's most awe-inspiring displays—the total solar eclipse. On this day, the Moon will align perfectly with the Sun and Earth, casting a shadow that will traverse a vast swath of North America, spanning from northern Mexico through the United States and into southeastern Canada.

The path of totality—the narrow corridor where the Moon completely obscures the Sun—will serve as the focal point for eclipse enthusiasts and scientists alike. Starting in the vicinity of Mazatlán, Mexico, the path will sweep across the continental United States, passing through states such as Texas, Arkansas, Missouri, Illinois, Indiana, Ohio, New York, and Vermont, before concluding in the Canadian provinces of Quebec and Newfoundland.

The duration of totality—the precious moments when the Sun's corona is visible—will vary along the path, reaching its maximum duration of approximately 4 minutes and 28 seconds near the town of Nazas in Mexico. As the eclipse progresses, observers located within the path of totality will be treated to varying durations of darkness, ranging from several minutes to just over two minutes, depending on their precise location within the shadow's trajectory.

Anticipation and Excitement Surrounding the Event

The forthcoming total solar eclipse has ignited a wave of anticipation and excitement among astronomers, eclipse chasers, and the general public alike. Months, if not years, of meticulous planning and preparation have culminated in a crescendo of anticipation as the event draws near.

For seasoned eclipse chasers, the allure of witnessing a total solar eclipse is irresistible. Many have traveled thousands of miles, braving challenging terrain and unpredictable weather conditions, in pursuit of the perfect vantage point from which to witness this cosmic spectacle. Whether they choose to observe from the rugged landscapes of the American Southwest or the bustling urban centers of the Northeast, eclipse chasers are united by their shared passion for experiencing nature's grandest theater.

Scientists and researchers are equally eager to capitalize on this rare opportunity to study the Sun's elusive corona. Armed with state-of-the-art telescopes, cameras, and instrumentation, they will endeavor to capture fleeting moments of totality in unprecedented detail, unraveling the mysteries of the solar atmosphere and advancing our understanding of solar physics.

The general public, too, is caught up in the excitement surrounding the eclipse, as communities across the eclipse path prepare to welcome throngs of visitors eager to witness this celestial spectacle firsthand. Festivals, educational programs, and public viewing events are being organized in cities and towns along the path of totality, offering residents and visitors alike the chance to come together and marvel at the wonders of the cosmos.

In conclusion, the 2024 total solar eclipse promises to be a momentous occasion, captivating audiences around the world with its breathtaking beauty and scientific significance. As the date approaches, anticipation continues to build, fueling a sense of wonder and excitement that transcends borders and unites humanity in the shared experience of witnessing one of nature's most extraordinary phenomena.

THE LONGEST TOTAL 1 SOLAR ECLIPSE IN THE U.S. SINCE 1806

As the anticipation builds for the upcoming total solar eclipse on April 8, 2024, one aspect of this celestial event stands out: it will be the longest total solar eclipse to grace the skies of the United States since 1806. In this section, we will explore the significance of this distinction, comparing it to past eclipses and delving into the duration of totality and its implications.

Comparison to Past Eclipses

The last time the United States experienced a total solar eclipse of comparable duration was over two centuries ago, on June 16, 1806. During that historic event, observers were treated to a spectacle that lasted up to 4 minutes and 55 seconds—a remarkable duration that captured the imagination of astronomers and the general public alike.

Known as "Tecumseh's Eclipse," in honor of the Shawnee leader Tecumseh, who purportedly foretold the eclipse as a sign of divine intervention, the 1806 eclipse left a lasting impression on those who witnessed it. From Baja California to Cape Cod, millions of Americans marveled at the sudden darkness that enveloped the landscape, prompting reflections on the mysteries of the cosmos and the fragility of human existence.

Fast forward to 2024, and once again, North America finds itself on the brink of another historic eclipse. While the duration of totality may not surpass the record set in 1806, it will come remarkably close, with observers in certain locations poised to experience up to 4 minutes and 28 seconds of darkness—a duration that rivals some of the longest eclipses in recent memory.

Duration of Totality and Its Significance

The duration of totality—the fleeting moments when the Moon completely obscures the Sun—holds immense significance for astronomers, eclipse chasers, and casual observers alike. During these precious minutes, the sky darkens, stars become visible, and the Sun's ethereal corona emerges in all its splendor, offering a rare glimpse into the outer reaches of our solar system.

For astronomers and researchers, the extended duration of totality presents a unique opportunity to study phenomena such as solar flares, prominences, and the Sun's magnetic field with unprecedented clarity. By capturing high-resolution images and spectroscopic data during the eclipse, scientists can unlock valuable insights into the dynamics of the Sun's atmosphere and advance our understanding of solar physics.

Eclipse chasers, too, are drawn to the allure of extended totality, as it allows for a more immersive and awe-inspiring experience of nature's grandest spectacle. Whether they choose to observe from remote mountaintops, expansive deserts, or bustling city centers, eclipse chasers relish the opportunity to witness the Sun's corona in all its glory, enveloped in the eerie darkness of totality.

Even casual observers cannot help but be captivated by the prolonged duration of totality, as it transforms the familiar landscape into a surreal tableau of light and shadow. From the eerie silence that descends upon the Earth to the breathtaking beauty of the solar corona, the experience of totality leaves an indelible impression on all who witness it, fostering a sense of wonder and humility in the face of the universe's majesty.

In conclusion, the 2024 total solar eclipse represents a rare convergence of celestial events, offering observers in North America the opportunity to witness one of the longest eclipses in recent memory. As the date approaches, anticipation continues to build, fueling excitement and wonder among astronomers, eclipse chasers, and the general public alike.

SOLAR MAXIMUM AND ITS IMPACT ON THE ECLIPSE

As the 2024 total solar eclipse approaches, astronomers and enthusiasts are eagerly anticipating not only the rare celestial event itself but also the unique circumstances surrounding it. One of the key factors shaping the eclipse experience is the phenomenon known as solar maximum. In this section, we will explore the concept of solar maximum and its profound impact on the appearance of the Sun's corona during the eclipse.

Explanation of Solar Maximum

Solar maximum is a phase in the 11-year solar cycle during which the Sun's magnetic activity reaches its peak. This period is characterized by an increase in the number and frequency of solar flares, coronal mass ejections (CMEs), and sunspots—dark areas on the Sun's surface where magnetic activity is concentrated. Solar maximum represents the culmination of the solar cycle, marking a peak in solar activity before gradually transitioning into a period of decreased activity known as solar minimum.

The solar cycle is driven by the Sun's magnetic field, which undergoes a process of reversal approximately every 11 years. As the Sun's magnetic field becomes increasingly tangled and complex during the buildup to solar maximum, it gives rise to intense magnetic activity on the Sun's surface, manifesting as sunspots and solar flares. These phenomena are visible from Earth and can have significant effects on our planet's space environment, including disruptions to satellite communications, power grids, and radio transmissions.

Influence on the Appearance of the Sun's Corona

During a total solar eclipse, the Sun's corona—the outermost layer of the Sun's atmosphere—becomes visible as the Moon blocks out the Sun's bright disk. The appearance of the corona is influenced by various factors, including the Sun's magnetic activity, which peaks during solar maximum.

At solar maximum, the Sun's corona tends to exhibit a more symmetrical and extended shape, with prominent streamers and loops extending outward from the Sun's surface. These features are the result of the Sun's magnetic field shaping and confining the hot, ionized gases of the corona. The increased magnetic activity during solar maximum can also lead to the formation of coronal holes—regions of reduced density and cooler temperatures in the corona—where solar wind flows more freely into space.

During a total solar eclipse occurring at solar maximum, observers may notice that the Sun's corona appears larger, more structured, and more dynamic than during eclipses occurring at other points in the solar cycle. The extended streamers and loops of the corona are more pronounced, creating a mesmerising display of solar activity against the darkened sky.

In addition to its visual impact, the heightened magnetic activity during solar maximum can also influence the behavior of the Sun's corona in other ways. For example, the presence of strong magnetic fields can accelerate the solar wind, leading to enhanced space weather effects on Earth and throughout the solar system.

In conclusion, solar maximum plays a crucial role in shaping the appearance of the Sun's corona during a total solar eclipse. As the 2024 eclipse approaches, astronomers are eager to study the dynamics of the corona during this peak period of solar activity, gaining valuable insights into the Sun's magnetic field and its influence on the space environment.

DARKER THAN EVER : THE DARKEST U.S. TOTAL SOLAR ECLIPSE IN 217 YEARS

As the countdown to the 2024 total solar eclipse continues, excitement mounts over the prospect of witnessing one of the darkest solar eclipses in recent memory. This forthcoming celestial event promises to plunge parts of North America into darkness, offering observers a rare opportunity to experience the wonders of the cosmos firsthand. In this section, we will explore the factors influencing the darkness during totality and the implications for viewing conditions during the eclipse.

Factors Affecting Darkness During Totality

Several factors contribute to the level of darkness experienced during totality, the brief period when the Moon completely obscures the Sun. These factors include the magnitude of the eclipse—the fraction of the Sun's diameter obscured by the Moon—and the width of the path of totality.

In the case of the 2024 total solar eclipse, the magnitude of the eclipse is expected to reach 1.05, meaning that over 100% of the Sun's diameter will be covered by the Moon at maximum eclipse. This high magnitude, combined with the relatively wide path of totality—approximately 115 miles—contributes to the darkness experienced during the eclipse.

Additionally, the clarity of the sky and the presence of atmospheric conditions such as clouds, haze, and pollution can also impact the darkness observed during totality. Clear, unpolluted skies tend to enhance the contrast between the darkened sky and the bright corona of the Sun, resulting in a more dramatic and immersive eclipse experience.

Implications for Viewing Conditions

The heightened darkness expected during the 2024 total solar eclipse has significant implications for viewing conditions and the overall eclipse experience. Observers located within the path of totality will be treated to a rare spectacle as daylight gives way to twilight-like conditions, with the Sun's corona shining brightly against the darkened sky.

For astronomers and researchers, the darkness during totality provides an ideal backdrop for studying the Sun's outer atmosphere and phenomena such as solar flares, prominences, and coronal mass ejections. By capturing high-resolution images and spectroscopic data during the eclipse, scientists can gain valuable insights into the dynamics of the Sun's atmosphere and advance our understanding of solar physics.

Eclipse chasers and enthusiasts, too, eagerly anticipate the heightened darkness of the 2024 eclipse, as it enhances the immersive nature of the experience. From the eerie silence that descends upon the landscape to the breathtaking beauty of the solar corona against the darkened sky, totality offers a mesmerizing glimpse into the wonders of the universe.

However, the darkness during totality also presents challenges for observers and photographers, particularly in terms of capturing the fleeting moments of the eclipse with precision and clarity. Proper planning, equipment, and techniques are essential for maximizing the viewing and photographic opportunities presented by the eclipse.

In conclusion, the 2024 total solar eclipse promises to be one of the darkest solar eclipses in recent memory, offering observers a rare opportunity to witness the wonders of the cosmos against the backdrop of a darkened sky. As the date of the eclipse approaches, anticipation continues to build, fueling excitement and wonder among astronomers, eclipse chasers, and the general public alike.

COMET SIGHTING : A RARE CELESTIAL PHENOMENON

As astronomers and skywatchers eagerly anticipate the 2024 total solar eclipse, there is the possibility of an additional celestial treat: the sighting of Comet 12P/Pons-Brooks. Comets, with their ethereal tails and unpredictable appearances, have long captivated humanity's imagination. In this section, we will explore the possibility of spotting Comet 12P/Pons-Brooks during the eclipse and the importance of comets in the context of eclipses.

Possibility of Spotting Comet 12P/Pons-Brooks

Comet 12P/Pons-Brooks, also known as the "Devil Comet," is a periodic comet with an orbital period of approximately 71 years. Discovered by astronomer

Jean-Louis Pons in 1812 and subsequently rediscovered by William Robert Brooks in 1883, this comet has a storied history of appearances throughout the centuries.

As the 2024 total solar eclipse approaches, astronomers have identified Comet 12P/Pons-Brooks as a potential candidate for observation during totality. The comet's projected position in the sky, combined with its brightness and proximity to the Sun, make it a tantalizing target for observers hoping to catch a glimpse of this rare celestial phenomenon.

While the visibility of Comet 12P/Pons-Brooks during the eclipse is not guaranteed, astronomers and amateur skywatchers alike are preparing to scan the skies for any signs of its presence. Given the comet's relatively close proximity to Jupiter and its potential for outbursts, there is a chance that it may brighten sufficiently to be visible to the naked eye or through binoculars during totality.

For those hoping to catch a glimpse of Comet 12P/Pons-Brooks during the eclipse, careful planning and preparation will be essential. Observers should familiarize themselves with the comet's predicted position and trajectory, as well as the optimal viewing conditions for spotting it in the night sky. Additionally, the use of telescopes or binoculars equipped with solar filters may enhance the chances of detecting the comet against the backdrop of the darkened sky during totality.

Importance of Comets in Eclipses

Comets have played a significant role in shaping our understanding of eclipses and their celestial significance throughout history. In ancient times, comets were often viewed as omens or portents of significant events, including eclipses, wars, and natural disasters. Their unpredictable appearances and ethereal tails sparked fear, awe, and speculation among ancient civilizations, leading to the development of myths, legends, and superstitions surrounding comets and their connection to eclipses.

From a scientific perspective, comets provide valuable insights into the composition, structure, and evolution of the solar system. Comets are composed of icy materials, dust, and organic compounds, making them cosmic time capsules that preserve information about the conditions present during the early formation of the solar system. By studying comets and their interactions with the Sun, scientists can gain insights into the processes that shaped the evolution of our planetary neighborhood over billions of years.

In the context of eclipses, comets offer an additional layer of intrigue and fascination for observers and researchers alike. The rare conjunction of a comet sighting with a total solar eclipse creates a unique opportunity to study two celestial phenomena simultaneously, shedding light on their interconnectedness and the dynamic nature of the cosmos.

In conclusion, the possibility of spotting Comet 12P/Pons-Brooks during the 2024 total solar eclipse adds an exciting dimension to an already captivating celestial event. Whether or not the comet makes an appearance, its potential presence serves as a reminder of the beauty, mystery, and complexity of the universe we inhabit. As astronomers and skywatchers prepare to witness the eclipse, they do so with a sense of anticipation and wonder, eager to explore the wonders of the cosmos and uncover the secrets of the celestial realm.

RECORD -BREAKING VIEWER NUMBERS

As the date of the 2024 total solar eclipse draws near, anticipation continues to mount among astronomers, eclipse chasers, and the general public alike. With its path of totality traversing a significant portion of North America, this celestial event is poised to attract record-breaking viewer numbers across the continent. In this section, we will explore the expected viewership for the 2024 total solar eclipse and compare it to previous eclipses to understand the magnitude of this historic event.

Expected Viewership Across North America

The 2024 total solar eclipse is expected to draw millions of viewers from across North America, making it one of the most-watched astronomical events in recent history. According to estimates, approximately 40 million people in the United States alone live within the path of totality, which spans parts of 15 states from northern Mexico to southeastern Canada. When accounting for additional viewers in Mexico and Canada, the total number of potential viewers exceeds 50 million, surpassing the viewership of previous eclipses by a wide margin.

In addition to those residing within the path of totality, millions more are expected to travel to witness the eclipse from nearby areas or even from other parts of the world. Eclipse chasers, in particular, are known for their willingness to travel long distances to experience totality, often planning years in advance to secure the perfect viewing spot. The allure of witnessing the longest total solar eclipse in the United States since 1806, combined with the rarity of the event and its significance in the astronomical community, is expected to drive unprecedented levels of interest and participation.

The widespread availability of information and resources online has also contributed to the expected surge in viewership for the 2024 eclipse. Websites, social media platforms, and streaming services will offer live coverage of the event, allowing viewers to experience the eclipse from the comfort of their own homes or offices. This accessibility, coupled with the inherent excitement and wonder surrounding a total solar eclipse, is likely to attract a diverse and global audience eager to witness this cosmic spectacle.

Comparison to Previous Eclipses

In comparison to previous total solar eclipses, the expected viewership for the 2024 eclipse is set to eclipse all records. The 2017 total solar eclipse, which traversed a narrow path across the United States from coast to coast, drew an estimated 215 million viewers, making it one of the most-watched events in American history. However, the 2024 eclipse is expected to surpass this figure significantly, thanks to its longer duration of totality, broader path of visibility, and heightened public awareness and interest.

Historically, total solar eclipses have captivated humanity's imagination and attracted large crowds of spectators eager to witness the rare spectacle. From ancient civilizations who interpreted eclipses as omens or messages from the gods to modern-day astronomers who study them for scientific research, eclipses have always held a special significance in human culture and society. The 2024 eclipse, with its potential to break viewing records and capture the attention of millions, continues this tradition and reaffirms the enduring fascination with celestial events that transcend borders and unite humanity in awe and wonder.

In conclusion, the 2024 total solar eclipse is poised to attract record-breaking viewer numbers across North America, surpassing previous eclipses in terms of both magnitude and significance. With millions of people expected to witness the event firsthand and millions more tuning in from around the world, the eclipse promises to be a historic and unforgettable experience for all who have the opportunity to witness it. As the date approaches, anticipation continues to build, fueling excitement and wonder among astronomers, eclipse chasers, and the general public alike.

URBAN ECLIPSE: OBSERVING FROM MAJOR CITIES

As the 2024 total solar eclipse approaches, millions of people across North America are preparing to witness this rare celestial event. While many eclipse chasers will venture to remote locations to experience totality, a significant number of observers will have the opportunity to witness the eclipse from major urban centers within the path of totality. In this section, we will explore the impact of urban areas on the viewing experience of the eclipse and highlight notable cities within the path of totality.

Impact of Urban Areas on Viewing Experience

Observing a total solar eclipse from a major city presents both opportunities and challenges for viewers. On one hand, urban areas offer convenient access to amenities, transportation, and infrastructure, making it easier for residents and visitors to access viewing locations and participate in eclipse-related events and activities. Additionally, the presence of skyscrapers and landmarks can provide unique vantage points for observing the eclipse and capturing memorable photographs of the event against the backdrop of the city skyline.

On the other hand, urban areas also pose obstacles to eclipse viewing, including light pollution, air pollution, and obstructed views caused by tall buildings and dense urban development. Light pollution, in particular, can significantly diminish the visibility of the eclipse, obscuring fainter celestial phenomena such as the solar corona and stars. Additionally, the presence of clouds and haze in urban areas can further obstruct views of the eclipse, reducing the overall quality of the viewing experience for observers.

Despite these challenges, many urban dwellers are eager to witness the eclipse from their own neighborhoods and communities, embracing the opportunity to share in this rare celestial event with friends, family, and fellow residents. By organizing viewing parties, educational events, and public gatherings, cities within the path of totality can foster a sense of community and camaraderie among eclipse enthusiasts, uniting people from diverse backgrounds in their shared fascination with the wonders of the cosmos.

Notable Cities within the Path of Totality

Several major cities in North America lie within the path of totality for the 2024 total solar eclipse, offering residents and visitors the chance to witness this extraordinary event without straying far from home. Among the notable cities within the path of totality are:

1. Dallas-Fort Worth-Arlington, Texas: The sprawling metropolitan area of Dallas-Fort Worth-Arlington is expected to experience over 4 minutes of totality during the eclipse, making it one of the longest viewing durations in the region.

2. Indianapolis, Indiana: Located in the heart of the Midwest, Indianapolis will be treated to approximately 3 minutes and 40 seconds of totality, providing residents with ample time to observe the eclipse and its celestial spectacle.

3. Cleveland, Ohio: Nestled along the shores of Lake Erie, Cleveland offers a picturesque backdrop for eclipse viewing, with over 3 minutes of totality expected to occur within the city limits.

4. Montreal, Canada: North of the border, the vibrant city of Montreal will experience partial totality during the eclipse, providing residents with a unique opportunity to witness this rare celestial event from an urban setting.

These cities, along with numerous others within the path of totality, are expected to attract large numbers of viewers eager to witness the eclipse firsthand and participate in eclipse-related festivities and events. From rooftop parties to public gatherings in city parks, urban areas offer a variety of options for experiencing the eclipse and making lasting memories with friends and family.

In conclusion, observing a total solar eclipse from a major city presents both challenges and opportunities for viewers. While urban areas may pose obstacles such as light pollution and obstructed views, they also offer convenient access to amenities and infrastructure, as well as unique vantage points for observing the eclipse. As the 2024 eclipse approaches, residents and visitors in cities within the path of totality are preparing to embrace this rare celestial event and share in the wonder and excitement of experiencing totality from an urban perspective.

UNIQUE ASPECTS OF THE 2024 ECLIPSE

The 2024 total solar eclipse is poised to be an extraordinary celestial event that captivates the attention of astronomers, eclipse enthusiasts, and the general public alike. While all total solar eclipses are inherently awe-inspiring, the 2024 eclipse stands out for several unique aspects that set it apart from previous eclipses. In this section, we will explore the additional features that make the 2024 eclipse special and its significance for astronomers and eclipse enthusiasts around the world.

Additional Features Setting This Eclipse Apart

1. Long Duration of Totality: The 2024 eclipse is expected to offer a particularly long duration of totality, with some areas

within the path of totality experiencing over 4 minutes of complete darkness. This extended period of totality provides observers with ample time to witness and study the various phenomena associated with the eclipse, including the solar corona, prominences, and shadow bands.

2. Solar Maximum Alignment: The 2024 eclipse coincides with a period of solar maximum, during which the Sun's magnetic activity is at its peak. This alignment enhances the appearance of the solar corona, making it more symmetrical and structured than during periods of low solar activity. Observers can expect to see dynamic features such as streamers, loops, and prominences emanating from the Sun's surface during totality.

3. Darkness and Visibility: The 2024 eclipse is projected to be one of the

darkest total solar eclipses in recent history, with a high magnitude and wide path of totality contributing to enhanced darkness during the event. This darkness allows for the visibility of fainter celestial objects such as planets and stars, providing a unique opportunity for observers to witness multiple astronomical phenomena in the same sky.

4. Comet Sighting Potential: One of the most exciting aspects of the 2024 eclipse is the possibility of spotting Comet 12P/Pons-Brooks during totality. While not guaranteed, the comet's proximity to Jupiter and potential for outbursts make it a tantalising target for observers hoping to capture a rare celestial event on camera.

5. Urban Observing Opportunities: Unlike many total solar eclipses that

traverse remote or inaccessible regions, the path of totality for the 2024 eclipse includes several major urban centers such as Dallas-Fort Worth, Indianapolis, and Cleveland. This presents a unique opportunity for residents of these cities to witness totality without the need for extensive travel, fostering widespread participation and public engagement with the eclipse.

Significance for Astronomers and Eclipse Enthusiasts

The unique aspects of the 2024 eclipse hold significant scientific and cultural significance for astronomers and eclipse enthusiasts worldwide. For astronomers, the eclipse offers a rare opportunity to study the dynamics of the solar corona and the Sun's outer atmosphere during a period of peak solar activity. By capturing high-resolution images and spectroscopic data during totality, scientists can gain valuable insights into the processes driving solar variability and space weather phenomena.

For eclipse enthusiasts, the 2024 eclipse represents a once-in-a-lifetime opportunity to witness the majesty and grandeur of a total solar eclipse in all its glory. The extended duration of totality, combined with the heightened darkness and potential for comet sightings, promises to create a truly unforgettable experience for observers lucky enough to be within the path of totality.

Moreover, the accessibility of the eclipse to urban populations highlights the universal appeal and significance of astronomical events in modern society. By bringing the wonder of the cosmos to city dwellers and urban communities, the 2024 eclipse serves as a reminder of the profound connection between humanity and the natural world, inspiring curiosity, awe, and appreciation for the beauty and complexity of the universe.

In conclusion, the 2024 total solar eclipse promises to be a remarkable and unforgettable event that showcases the wonders of the cosmos in all their splendor. With its long duration of totality, alignment with solar maximum, potential for comet sightings, and accessibility to urban populations, this eclipse offers a unique opportunity for scientific discovery, cultural enrichment, and personal reflection. As the date of the eclipse approaches, anticipation continues to build, fueling excitement and wonder among astronomers, eclipse enthusiasts, and the general public alike.

PREPARATION AND SAFETY GUIDELINES FOR VIEWING THE 2024 SOLAR ECLIPSE

As the highly anticipated 2024 total solar eclipse approaches, it's crucial for observers to prepare adequately and prioritize safety while witnessing this rare celestial event. Viewing a solar eclipse requires caution and proper equipment to protect your eyes from the harmful effects of the Sun's intense rays. In this section, we will discuss essential preparation and safety guidelines to ensure a safe and enjoyable eclipse viewing experience.

*How to Safely View a Solar Eclipse***

1. Use Solar Viewing Glasses: The most important safety precaution when viewing a solar eclipse is to wear certified

solar viewing glasses. These glasses are specially designed to block harmful ultraviolet, visible, and infrared radiation from the Sun while allowing you to see the eclipse safely. Ensure that your glasses are certified by reputable manufacturers and have the ISO 12312-2 certification printed on them.

2. Do Not Look Directly at the Sun: Never look directly at the Sun with your naked eyes, even during a partial eclipse. Doing so can cause permanent damage to your eyes, including solar retinopathy and blindness. Instead, use solar viewing glasses or other indirect viewing methods to observe the eclipse safely.

3. Use Solar Filters for Telescopes and Binoculars: If you plan to use telescopes or binoculars to observe the eclipse, make sure to equip them with appropriate solar filters. Solar filters

should be attached securely to the front aperture of the telescope or binoculars to block the Sun's intense light and prevent eye injury.

4. Create a Pinhole Projector: Another safe and easy way to view a solar eclipse is to create a pinhole projector. This simple device projects an image of the Sun onto a surface, allowing you to observe the eclipse indirectly. To make a pinhole projector, punch a small hole in a piece of cardboard or paper and hold it up to the Sun, allowing the sunlight to pass through the hole and project onto a surface below.

5. Use Welder's Glass: Welder's glass with a shade rating of at least 14 is another option for safe solar viewing. Welder's glass can be used as an alternative to solar viewing glasses or as an additional layer of protection when

observing the eclipse with telescopes or binoculars.

Recommended Equipment and Precautions

1. Plan Ahead: Before the day of the eclipse, familiarize yourself with the path of totality and choose a safe and accessible viewing location. Arrive early to secure your spot and avoid last-minute crowds and traffic.

2. Bring Adequate Supplies: In addition to solar viewing glasses or other viewing equipment, be sure to bring sunscreen, water, snacks, and comfortable clothing to stay hydrated and protected from the elements during the eclipse.

3. Protect Your Camera Equipment: If you plan to photograph the eclipse, use solar filters or eclipse glasses to protect

your camera lenses and sensors from damage. Never point your camera directly at the Sun without proper protection.

4. Be Mindful of Wildlife: If you're observing the eclipse in a natural or remote area, be respectful of wildlife and their habitats. Avoid disturbing animals or causing unnecessary noise and commotion during the eclipse.

5. Follow Local Regulations: Be aware of any local regulations or restrictions regarding eclipse viewing, especially in public parks, recreational areas, or private properties. Respect designated viewing areas and adhere to safety guidelines provided by local authorities and event organizers.

By following these preparation and safety guidelines, you can ensure a safe and memorable viewing experience of the 2024 total solar eclipse. Remember to prioritize safety at all times and take necessary precautions to protect your eyes and equipment from the Sun's intense radiation. With proper planning and adherence to safety protocols, you can enjoy the wonder and beauty of this rare celestial event without putting yourself or others at risk.

CAPTURING THE ECLIPSE: PHOTOGRAPHY TIPS AND TECHNIQUES

Photographing a total solar eclipse is an exhilarating experience that requires careful planning, the right equipment, and a solid understanding of photography techniques. Whether you're a seasoned photographer or a novice enthusiast, capturing the beauty and drama of a solar eclipse requires preparation and skill. In this section, we will explore essential photography tips and techniques for capturing the 2024 total solar eclipse.

Advice for Photographing the Eclipse

1. Research and Plan Ahead: Before the day of the eclipse, research the path of totality and choose a suitable location

with a clear view of the Sun. Consider factors such as weather conditions, accessibility, and potential obstacles that may affect your photography. Plan your shooting location and composition in advance to maximize your chances of capturing compelling images of the eclipse.

2. Use the Right Equipment: To photograph a solar eclipse, you'll need a digital camera with manual exposure settings, a sturdy tripod, and a telephoto lens with a focal length of at least 300mm to capture detailed images of the Sun. Additionally, you'll need solar filters or eclipse glasses to protect your eyes and camera equipment from the Sun's intense light.

3. Practice Safe Solar Photography: Never look directly at the Sun through your camera's viewfinder or LCD screen,

as doing so can cause permanent eye damage. Instead, use live view mode or a solar filter to compose and focus your shots safely. Attach solar filters to your lens or camera body to reduce the intensity of the Sun's light and prevent overexposure in your images.

4. Experiment with Exposure Settings: When photographing a solar eclipse, it's essential to adjust your camera's exposure settings to achieve the correct exposure for the Sun. Start with a low ISO setting (ISO 100 or lower) to minimize noise in your images, and use a narrow aperture (f/8 to f/16) to maximize depth of field and sharpness. Adjust your shutter speed to achieve a properly exposed image of the Sun, taking care not to overexpose or underexpose the solar disk.

5. Bracket Your Exposures: To ensure you capture the full range of tones and details in the eclipse, consider bracketing your exposures by taking multiple shots at different exposure settings. This technique allows you to capture highlights, shadows, and midtones separately and blend them together in post-processing to create a well-exposed final image.

Equipment Recommendations and Settings

1. Camera Body: Use a digital SLR or mirrorless camera with manual exposure settings for greater control over your images. Choose a camera with a high-resolution sensor and low noise performance to capture detailed and clear images of the eclipse.

2. Lens: Select a telephoto lens with a focal length of at least 300mm to capture close-up shots of the Sun and reveal details such as sunspots, prominences, and the solar corona. Consider using a teleconverter or zoom lens with a longer focal length for even greater magnification and detail.

3. Solar Filters: Invest in high-quality solar filters or eclipse glasses to protect your eyes and camera equipment from the Sun's intense light. Solar filters should be attached securely to your lens or camera body to block harmful ultraviolet, visible, and infrared radiation while allowing you to capture clear and sharp images of the eclipse.

4. Tripod: Use a sturdy tripod to stabilize your camera and minimize camera shake during long exposures. Choose a tripod with adjustable legs and a robust

mounting plate to support the weight of your camera and lens setup securely.

5. Settings: Set your camera to manual mode and adjust your exposure settings based on the brightness of the Sun. Start with a low ISO setting (ISO 100 or lower), a narrow aperture (f/8 to f/16), and a fast shutter speed (1/1000s or faster) to capture sharp and well-exposed images of the Sun. Use the camera's histogram and exposure meter to monitor your exposure levels and make adjustments as needed.

In conclusion, photographing a total solar eclipse is a challenging yet rewarding endeavor that requires careful planning, the right equipment, and technical expertise. By following these photography tips and techniques, you can capture stunning images of the 2024 eclipse and preserve the beauty and wonder of this rare celestial event for years to come. Remember to prioritize safety at all times and practice responsible solar photography to protect your eyes and equipment from the Sun's intense radiation. With practice and perseverance, you can create breathtaking images that showcase the magic and majesty of a total solar eclipse in all its glory.

www.ingramcontent.com/pod-product-compliance
Lightning Source LLC
Chambersburg PA
CBHW070438290526
45791CB00005B/2033